flavouring with **Olive Oil**

flavouring with

Olive Oil

Clare Gordon-Smith

photography by

James Merrell

Whitecap Books
Vancouver/Toronto

Art Director **Jacqui Small**

Art Editor **Penny Stock**

Design Assistant **Mark Latter**

Editor **Elsa Petersen-Schepelern**

Photography **James Merrell**

Food Stylist **Clare Gordon-Smith**

Stylist **Sue Skeen**

Production Consultant **Vincent Smith**

Our thanks to Christine Walsh and Ian Bartlett, Jenny Merrell, Peter Bray and Robert Roseman.

First published in Great Britain in 1996
by Ryland Peters & Small
Cavendish House, 51-55 Mortimer Street, London W1N 7TD

This edition published by Whitecap Books Ltd.
351 Lynn Avenue, North Vancouver,
British Columbia V7J 2C4

Produced by Mandarin Offset
Printed and bound in China

ISBN 1-55110-501-2

Notes:
Ovens should be preheated to the specified temperature. If using a convection oven, adjust times and temperature according to manufacturer's instructions.

All pickles and preserves should be processed in a boiling water-bath canner according to USDA guidelines.

The FDA has raised concerns about potential health hazards of homemade flavored oils. The problem occurs when bacteria that exists naturally in some flavoring items are introduced to the airless environment of oil. If you choose to make flavored oil, the following steps are recommended:

- Gently heat flavoring ingredients, such as crushed garlic and herbs, in the oil. Do not allow items to stand in unheated oil.
- Make small batches of flavored oil and use within 24 hours. Keep flavored oil refrigerated when not in use.

Olive oil is one of those rare commodities—something that tastes wonderful and is actually good for you. Specialist food shops conduct olive oil tastings, like wine tastings. The oils are poured into small white dishes and accompanied by pieces of bread. The oils are tasted by dipping the bread into each oil and comparing the flavors. Oils at such **tastings** are offered labeled according to their country or region of origin, and also to their grade. The grades of olive oil are determined by the method of harvesting and pressing. The most desirable in terms of quality are hand-harvested, first pressed, and cold pressed. The finest oils from each region are produced in this way. The best oil is **extra-virgin**, with an acidity of less than one percent. It is best used as a flavoring agent—in dressings, or poured over finished dishes—so its inherent flavors are not altered by heat. Other grades include fine virgin olive oil, semi-fine olive oil, refined olive oil, and pure olive oil. You should choose a lower grade oil for cooking and for making mayonnaise. However, **aioli**, the rich garlic mayonnaise of Provence, should always be made from good, green, extra-virgin olive oil.

the flavors of
Olive Oil

Olives are grown all around the Mediterranean, as well as in South Africa, Australia, and California. The top oil producers are Spain, Italy, Greece, and France. Each country's oil has a distinctive **flavor**, whose nuances and color vary further by region—affected by climate, soil, and fruit variety. French oil is very **sweet** and, though produced in small quantities, is of fine quality. Greek oil is more **aromatic**, and oils from the Pelopponese and Crete are very highly regarded. Spain is the largest producer of oil and its oil tastes very **fruity**.

Fine Italian oils are produced in Liguria, Tuscany, Umbria, and Apulia. Tuscan oils have a **peppery** aftertaste, which varies from district to district.

Californian olive oil is fruity and **mild**, and produced under varietal names such as Manzanilla, Sevillano, and Mission.

Australian oils are labeled, like their **wines**, with fruit varieties, climate, aspect of the groves, and pedigrees of the trees.

Cyprus produces some superb oils, while oil produced in Israel, the Lebanon, Turkey, Syria, Tunisia, Portugal, Morocco, Algeria, and South Africa is mostly consumed in the home market.

Appetizers

Chilled tomato soup
with basil pesto

A soup for high summer when tomatoes are at their reddest and sweetest. Serve with garlic croutons—don't bother to cut them into neat dice—just tear the bread roughly into pieces, sauté with garlic in olive oil until lightly golden, then drain.

Place the tomatoes in a blender or food processor and puree to a mush. Pass through a stainless steel strainer into a bowl, pushing through as much juice and pulp as possible, giving a creamy consistency. Thin if necessary with the olive oil and tomato juice. Mix in all the remaining ingredients, place the bowl inside a larger bowl, and fill the space between the two bowls with ice. Pour cold water over the ice and chill in the refrigerator for about 2 hours to develop the flavors.

Serve in soup bowls, garnished with the basil pesto, strips of roasted yellow bell peppers, a drizzle of extra-virgin olive oil and an ice cube, together with a bowl of garlic croutons.

2 lb. very ripe plum tomatoes

3 tablespoons olive oil

⅔ cup tomato juice

3-inch piece of cucumber, finely chopped

1 yellow bell pepper, roasted, peeled, seeded, and chopped

1 small red onion, finely chopped

1 tablespoon balsamic vinegar

2 tablespoons torn fresh basil leaves

salt and freshly ground black pepper

to serve

basil pesto (see page 21)

roasted yellow bell peppers

extra-virgin olive oil

ice cubes

garlic croutons

Serves 4

Tuscan bean soup
with green pesto

A rustic soup, with a delicious green pesto swirled through just before serving. If fava beans are out of season, substitute green flageolet beans or more white cannellini beans. If you're short of time, you could also use canned cannellinis or flageolets.

½ cup dried white cannellini beans

⅔ cup pure olive oil

1 onion, finely chopped

¼ cabbage, finely sliced

4 cups vegetable stock

1 cup fava beans (about 1 lb. in the pod)

2 zucchini, sliced and quartered

2 tablespoons snipped fresh oregano leaves

salt and freshly ground black pepper, to taste

to serve

4 tablespoons basil pesto (see page 21)

4 sprigs of basil (optional)

4 tablespoons crumbled fresh Parmesan cheese

Serves 4

To prepare the dried beans, soak overnight in cold water. Next day, drain, place in a large saucepan with cold water to cover, bring to a boil, and simmer gently until the beans are tender (about 30 minutes, depending on the age of the beans). Add a little salt about 5 minutes before the beans are done.

To make the soup, heat the olive oil in a large saucepan, add the onion and cook over a medium heat until softened and transparent. Add the cabbage and cook for a few minutes. Add the stock, bring to a boil and simmer, covered, for about 20 minutes. Add the fava beans and zucchini, and simmer for a further 10 minutes. Stir in the oregano and season with salt and black pepper.

Add the cannellini beans, with their cooking liquid, and simmer together for about 5 minutes.

To serve, divide the soup between 4 large heated soup bowls, then place a tablespoon of pesto on top. Add a sprig of basil, if using, and sprinkle with the roughly crumbled Parmesan.

a rustic Italian soup, with brilliant green basil pesto

Goat cheese
tomato and lentil salad

Bordeaux lentils have a nutty taste and stay whole and firm when cooked—their flavor contrasts well with the strong goat cheese. Puy lentils, darker in color, are more widely available, with a more robust flavour.

Place the lentils in a saucepan and cover with cold water. Add the garlic and half the parsley, bring to a boil and simmer for 10 minutes.

Add the onion, celery, and a pinch of salt, and simmer for 10 to 15 minutes more, adding a little hot water if required, until the lentils are just cooked but still firm. Discard the herbs and garlic, then toss the lentils in olive oil while still warm.

Crumble the goat cheese and mix into the lentils. Add the snipped chives, lemon juice, and the remaining parsley, chopped, then season to taste.

To serve, place salad leaves on each plate, spoon over the lentil and cheese mixture, then garnish with the cherry tomatoes, chopped into chunks.

½ cup Bordeaux, Puy, or other brown lentils

1–3 garlic cloves

1 bunch of fresh flat-leaf parsley

1 red onion, finely diced

1 celery stalk, finely diced

pinch of salt

1 cup virgin olive oil

8 oz. goat cheese

1 bunch of chives, snipped

¼ cup lemon juice

salt and freshly ground black pepper

to serve

salad leaves, such as arugula, spinach, watercress, or mâche (lamb's lettuce)

8 oz. cherry tomatoes

Serves 4

a very **ea**
with truffle oil to give th

Leek and new potato salad
with truffle oil

Truffle oil is the most important ingredient in this recipe, and is also delicious drizzled over pasta. Buy it in small quantities, such as the little bottles of white and black truffle oils pictured. Tiny baguette leeks look great in salads, but if you can't find baguettes, use larger leeks cut into chunks and leave the chunks intact when serving.

1½ lb. new potatoes

1 lb. young leeks (preferably baguettes)

2–4 tablespoons roughly chopped fresh flat-leaf parsley

3 tablespoons olive oil

sea salt and freshly ground black pepper

4 tablespoons truffle oil, to serve

Serves 4

Cook the potatoes in boiling salted water for about 20 minutes, until tender.

Trim the leeks and slice into 1-inch chunks, if large. Wash thoroughly and cook in boiling salted water for 5 to 7 minutes, until just tender.

Place the parsley, oil, salt, and pepper in a bowl. Drain the potatoes, slice thickly, and add to the dressing while still warm, so they absorb the flavors. Drain the leeks thoroughly and add to the potatoes. Stir well, taste, and adjust the seasoning. Spoon onto serving plates, then drizzle with truffle oil, and serve.

t spectacular salad—

ra **dash of luxury**

Fried quail egg salad
with bacon and leaves

For the best look, flavor, and texture, fry the eggs, but boiled eggs may also be used. Nasturtium flowers make a colorful, peppery addition to this salad in summer.

mixed leaf salad

nasturtium flowers (optional)

4 tablespoons olive oil

8 quail eggs

4 slices bacon, cut crosswise into ½ inch strips

2 tablespoons red wine vinegar

1 teaspoon wholegrain mustard

salt and freshly ground black pepper

Serves 4

Arrange the leaves (and flowers if using) on 4 appetizer plates.

Heat 1 tablespoon of the oil in a small skillet and lightly cook 4 of the eggs. Carefully place on the leaves. Add another tablespoon of oil and repeat with the remaining eggs.

Add the bacon pieces to the skillet, sauté until crisp, then add to the salad.

Whisk in the remaining 2 tablespoons of olive oil, together with the vinegar and mustard, then pour over the eggs and salad leaves. Season with salt and freshly ground black pepper

Serve with crusty bread.

Bruschetta
with tomatoes and olives

Bruschettas make perfect fast snacks—and also delicious appetizers for a summer lunch. Try other toppings too, such as the goat cheese on page 29, the artichokes on page 26, or the eggplant on page 58.

Split the focaccias in half and toast lightly. Spread with a dollop of green basil pesto, then add the tomatoes and olives. Add a slice of mozzarella to each, then drizzle with olive oil and place under the broiler for a few minutes. Serve, garnished with sprigs of basil leaves.

4 focaccias or sliced baguette loaf

green basil pesto (see opposite)

6 oz. vine-ripened cherry tomatoes

⅓ cup black olives, pitted and quartered

2–4 mozzarella cheeses, about 4 oz. each, sliced

Italian single estate olive oil

sprigs of basil, to garnish

Serves 4

Olive tapenade

If available, use olives and olive oil from the same region or country of origin.

Pit and chop the olives, then place in a blender or food processor with the capers, anchovies, and garlic, and blend thoroughly.
Add the olive oil gradually until the mixture reaches a spreadable consistency.
Slice and toast the baguette, spread with the tapenade, and serve with broiled tomato slices.

6 oz. black olives

2 tablespoons capers

4 anchovy fillets, rinsed and chopped

2 garlic cloves, crushed

⅔ cup olive oil

to serve

1 baguette loaf

broiled tomato slices

Serves 4

Green basil pesto

4 oz. fresh basil leaves

2 tablespoons pignoli nuts

¼ cup fresh Parmesan cheese, grated

⅔ cup olive oil

salt

Makes one ½-pint jar

Ready-made pesto is widely available in shops and supermarkets, but it's never as good as homemade. Whenever you see some good fresh basil, or if you grow your own, try this recipe—it's a revelation!

Place all the ingredients in a blender or food processor and puree to a thick cream, adding a little more oil if it is too thick. Add salt to taste. Spoon into a jar, and float a thin layer of oil on the top. Store in the refrigerator until ready to use.

Flavored oil

virgin olive oil

flavorings

a choice of:
1 sprig of rosemary,
2 sprigs of thyme,
1 tablespoon black peppercorns,
2 sprigs of bay leaves,
2 pieces of lemon peel
or 1 tablespoon dried chiles

You can buy flavored oils, but it's simple to make your own. Make very small quantities, store them in the refrigerator and use them quickly. Sterilize both bottles and corks thoroughly before using. To avoid possible bacterial growth, it is best not to make garlic or truffle oils yourself.

If using herbs, wash them thoroughly before use. Pour the oil into a saucepan and gently heat together with the herbs or other flavorings. Pour the hot oil into sterilized bottles, then add the heated herbs or flavorings. Store in the refrigerator and use within 24 hours.*

*Warning: please refer to FDA directive on page 4.

Left, front row, from left, Bruschettas with marinated artichokes (recipe page 26), with olive tapenade (page 20), and with green basil pesto (page 21). Right, with green basil pesto, and with tomatoes and olives (recipe page 20—see also back jacket).

Char-grilled asparagus
with a warm bread salad

The Italians use bread soaked in oil or vinegar as a base for sauces and soups. Char-grilling or roasting asparagus keeps the spears crisp on the inside. Caperberries are delicious if you can find them—but ordinary capers also work well.

1 lb. asparagus

oil, for brushing

2 teaspoons balsamic vinegar, to serve

warm bread salad

3 oz. bread (sourdough or country style)

⅔ cup white wine vinegar

1 hard-boiled egg, quartered

1 bunch of fresh flat-leaf parsley, roughly chopped

2 oz. caperberries, (if available), or capers packed in salt or brine

⅔ cup virgin olive oil

salt and freshly ground black pepper

Serves 4

To make the salad, tear the bread into chunks and mix with half the vinegar until softened. Squeeze dry and mix with the egg, parsley, caperberries or capers, salt, and pepper. Heat the oil and remaining wine vinegar in a saucepan until warm, then stir in the bread mixture. Keep the mixture warm while you cook the asparagus.

Heat a grill pan on top of the stove and lightly brush with olive oil. Add the asparagus spears and cook on all sides until slightly shriveled and brown. Alternatively, place in a roasting pan and cook in a preheated oven at 400°F for 5 to 7 minutes until browned. Place on a warmed serving plate, spoon over the warm bread salad, sprinkle with the balsamic vinegar, and serve immediately.

Baby artichokes
marinated in olive oil and lemon

These wonderfully flavored artichokes can be made in advance and stored in sealed jars. Serve them in this salad, as a bruschetta topping, or as an antipasto with sliced salami.

Peel away the outer layer of the artichoke stems and trim the leaves with a knife or scissors. (Young, tender artichokes have no prickly chokes.) Brush over all the cut surfaces with a little lemon juice to prevent browning. Place in a shallow saucepan with a tightly fitting lid. Add the lemon juice, bay leaves, peppercorns, parsley, olive oil, sea salt, and water to cover. Bring to a boil and simmer for 10 to 15 minutes until tender. Set aside to marinate for at least 24 hours. When ready to serve, drain and reserve the liquid to use again as a vinaigrette. Cut the artichokes in half, bunch the spinach or mustard leaves on the toasted bread, place the artichoke halves on top and pour over some of the marinade.

2 lb. small artichokes

juice of 3 lemons

3 bay leaves

10 black peppercorns

1 bunch of fresh flat-leaf parsley, chopped

⅔ cup olive oil

sea salt

to serve

8 oz. baby spinach or mustard leaves

4 slices sourdough bread, toasted

Serves 4

serve as an **antipasto**, on bruschetta, or as a salad with crusty bread

Goat cheese
marinated in olive oil

Marinate your own goat cheese in olive oil—it can then be used broiled on toast, as a pasta sauce with snipped herbs, or served with a selection of crisp crackers.

1 lb. goat cheese, sliced and quartered

2½ cups virgin olive oil

½ cup pitted black olives (optional)

4 garlic cloves

a selection of:
1 small red chile,
2–3 sprigs of thyme or rosemary,
2 bay leaves,
and 1 tablespoon peppercorns

lemon peel

sea salt

Makes two ½-pint jars

Sterilize 2 small Mason jars by placing in a deep saucepan, pouring over boiling water and simmering for 15 minutes. When cool, place the goat cheese in the sterilized jars. Pour in the olive oil, add salt, then push in the olives, if using. Blanch the garlic cloves, herbs, spices, and lemon peel in boiling water for 1 minute, drain and add to the jar. Cover and leave to marinate in the refrigerator for 2 days before using, then use within 1 week.

Mushrooms and anchovies
with sautéed polenta wedges

great for lunc

Polenta can be bought ready made or prepared according to the recipe below. As an alternative, use slices of focaccia, drizzled with olive oil and baked in the oven until crispy. You can also make this recipe without the anchovies.

To make the polenta, place in a pan, add the boiling water, and prepare according to the package instructions. When done, spread into a deep pan to set (about 5 minutes), then cut into wedges and sauté in the olive oil until lightly browned. Alternatively, place the pieces in a baking pan, drizzle with a little olive oil, and cook in a preheated oven at 400°F for about 10 minutes.
Rinse the porcini, soak in warm water for 20 minutes, then drain on paper towels, and roughly chop. Reserve the soaking liquid.
Heat the oil in a skillet, and sauté the garlic for a few minutes until softened and transparent.
Add the porcini and sauté for a few more minutes. Stir in the sherry, chopped anchovies, and porcini soaking liquid, bring to a boil, and simmer for about 5 minutes. Taste and adjust the seasoning.
Remove the mushrooms, then bring the sauce to a boil and reduce to about 4 to 6 tablespoons.
To serve, spoon the mixture onto the warm polenta, pour over the sauce, and sprinkle with oregano.

as an appetizer—easy to make
nd absolutely **packed with flavor**

8 polenta pieces or homemade polenta (see below)

about 4 oz. dried porcini mushrooms

3 tablespoons virgin olive oil

1 garlic clove, crushed

1 tablespoon dry sherry

2 oz. anchovies, canned in olive oil, finely chopped

sea salt and freshly ground black pepper

chopped fresh oregano leaves, to serve

fried polenta

1 cup pre-cooked polenta

1 cup boiling water

olive oil, for sautéing

Serves 4

Entrees

Char-grilled tuna
with red onion salad and gremolata

Marinating is a way of giving the fish extra flavor. This recipe is perfect for cooking on a barbecue in the summer, but you can get a similar effect using a grill pan on top of the stove. This recipe is also good with other big fish such as swordfish.

To make the gremolata, mix the parsley with the olive oil, lemon juice and zest, salt, and freshly ground black pepper. Set aside.

To make the salad, place the sliced onions in a shallow dish, pour over the olive oil and balsamic vinegar, and sprinkle with the herbs and seasonings. Set aside to develop the flavors.

Brush the tuna steaks with the olive oil. Heat a grill pan on top of the stove and, when very hot, quickly sear the tuna on both sides. Add seasoning, then continue to cook for about 5 minutes.

Divide the onion salad between 4 dinner plates with the tuna beside. Place a spoonful of gremolata on each steak, and serve with wedges of lemon.

4 tuna steaks

⅔ cup olive oil

salt and freshly ground black pepper

lemon wedges, to serve

gremolata

1 large bunch of fresh flat-leaf parsley, roughly chopped

½ cup olive oil

juice and finely sliced or grated zest of 1 lemon

salt and freshly ground black pepper

red onion salad

3 red onions, sliced

2 tablespoons extra-virgin olive oil

1 tablespoon balsamic vinegar

3 tablespoons finely chopped fresh flat-leaf parsley

salt and freshly ground black pepper

Serves 4

Roasted cod
with olive tapenade

The strong flavor of this tapenade works very well with cod or other big fish that hold their shape, such as haddock or flounder.

4 cod steaks, about 4 oz. each

⅔ cup pitted green olives

1 tablespoon capers packed in salt, or in brine (drained)

1 bunch of fresh flat-leaf parsley, chopped

8 anchovy fillets

3 tablespoons olive oil

sea salt and freshly ground black pepper

steamed couscous, to serve

Serves 4

Place the cod steaks in a roasting pan.
Mix the olives with the capers and parsley.
Spread this mixture on top of the cod steaks, place 2 anchovy fillets in a criss-cross on top of each steak, pour over the olive oil, and season with salt and freshly ground black pepper.
Cook in a preheated oven at 400°F for about 15 to 20 minutes until the fish turns white and opaque.
Serve with steamed couscous.

Seared beef fillet
with olives, tomatoes, and arugula

A spectacularly easy recipe, packed with flavor and goodness. Cook the beef in a grill pan on top of the stove, or very quickly in a searingly hot oven.

1 lb. beef fillet

4 garlic cloves, sliced lengthwise

⅔ cup olive oil

4 bay leaves

sea salt and freshly ground black pepper

to serve

1 large bunch of fresh arugula

4 ripe plum tomatoes, sliced

handful of black olives, preferably *herbes de provence* (optional)

extra-virgin olive oil, preferably from Provence

1 tablespoon torn fresh flat-leaf parsley

Serves 4–6

Cut slits into the surface of the beef and insert the garlic slices. Pour over the olive oil and sprinkle with freshly ground black pepper. Tuck the bay leaves underneath and set aside for a few minutes.

To prepare the arugula salad, place the leaves in a bowl, add the tomatoes and olives, if using. Drizzle with the extra-virgin olive oil.

Heat a grill pan on top of the stove until very hot. Add the beef and bay leaves and cook the beef for about 5 minutes on each side, depending on how rare you like your beef.

If roasting in the oven, cook in a preheated oven at 400°F for 10 minutes for rare beef, 15 minutes for medium, and 20 minutes for well done.

Discard the bay leaves. Set the beef aside, covered, for 5 minutes, to allow the meat to rest, then cut into thick slices.

Place a pile of arugula leaves on each plate, top with the beef slices, add the sliced tomatoes and black olives, if using, then dress with a little more extra-virgin olive oil and sprinkle with torn fresh flat-leaf parsley and sea salt.

Provençal lamb daube

This is a quick and easy update of one of the great classics from the South of France. Use oil from Provence, if you can find it, for a truly authentic flavor.

Place the lamb in a shallow pan.

Mix the marinade ingredients together and pour over the meat. Place in the refrigerator overnight or up to 2 days, depending on time available. Remove from the refrigerator about 1 hour before cooking to return to room temperature.

Remove the lamb from the marinade and pat dry with paper towels. Reserve the marinade.

Heat the olive oil in a large skillet and brown the meat on all sides.

Transfer to a casserole dish, sprinkle the flour over the oil and juices in the skillet, stir and cook for about 1 minute, then add the brandy.

Bring to a boil, remove from the heat, and pour over the meat in the casserole dish.

Pour in the marinade and stock, adding water to cover the meat if necessary, season well, and bring up to boiling point on top of the stove.

Transfer to a preheated oven and cook slowly at 325°F for 1 to 1½ hours.

Serve the daube with stewed flageolet beans and mashed potato made with olive oil.

6 large lamb shanks, left whole, or 1 leg of lamb, about 3 lb., cut into thick slices

3 tablespoons olive oil

2 tablespoons flour

½ cup brandy

1¼ cups vegetable or lamb stock

salt and freshly ground black pepper

marinade

4 garlic cloves

2 carrots

1¼ cups red wine

2 sprigs of thyme

2 sprigs of parsley

2 strips of orange peel

4 tablespoons olive oil

Serves 6

Marinated chicken
with avocado and spinach salad

It is important to use free-range chicken to give a good flavor in such a simple recipe—or in any other! Serve it either hot or cold, with a salad such as this one.

Place the chicken in a single layer in a roasting pan, mix the remaining ingredients together and pour over the chicken. Marinate in the refrigerator for at least 30 minutes, or longer if possible.

Cook in a preheated oven at 400°F for 20 minutes, or until the juices run clear when the chicken is pierced with a skewer through the thickest part.

To make the salad, blanch the beans in boiling water and refresh in iced water. Using a teaspoon, scoop out balls of avocado and place in a bowl with the beans, lettuce, and spinach leaves.

Whisk the oil with the vinegar, salt, and freshly ground black pepper, pour over the salad and serve immediately with the chicken.

A dish of boiled new potatoes would be a suitable accompaniment.

6 pieces of free-range chicken, about 2 lb.

juice of 1 lemon

½ cup virgin olive oil

½ cup pitted green olives

1 tablespoon capers, packed in salt

1 bay leaf

4 sprigs of rosemary

avocado and spinach salad

4 oz. green beans

1 ripe avocado

1 escarole or other soft lettuce

4 oz. baby spinach leaves

⅔ cup extra-virgin olive oil

¼ cup balsamic vinegar

salt and freshly ground black pepper

Serves 4–6

an easy dish for a midweek dinner party

Chicken breasts
with pancetta and rosemary oil

Rosemary oil is a wonderful flavoring ingredient. Use it for sautéing, roasting, or as a quick dressing for salads and steamed vegetables. Serve this dish with whole new potatoes, tossed in olive oil and crushed garlic, sprinkled with sea salt, then baked in the oven at the same time as the chicken.

4 skinless chicken breasts, free-range if available

4 slices Smithfield ham, or other high-quality ham

rosemary oil

½ cup extra-virgin olive oil

6 sprigs of rosemary

3 whole garlic cloves

pancetta filling

4 oz. pancetta, finely chopped

½ cup ricotta cheese

sea salt and freshly ground black pepper

Serves 4

To make the rosemary oil, place all the ingredients in a small saucepan, heat gently until tepid, then set aside for about 15 minutes until ready to use.

To make the filling, mix the pancetta with the ricotta, salt, and freshly ground black pepper.

Using a small, sharp knife, cut pockets down the sides of the chicken breasts and loosely pack with the filling. Wrap each breast in a slice of ham.

Place the chicken in a roasting pan, pour over the rosemary oil, and bake in a preheated oven at 400°F for 20 minutes.

Serve immediately with oven-baked garlic potatoes and a green leafy salad.

Braised pheasant
with apples, fennel, and sage

A winter dish which also works well with chicken, guinea fowl, quail, or other birds. Omit the fennel if it is unavailable—the dish is equally good without it. You could also leave out the flour—the golden sauce will be thinner, but it will still taste terrific.

Heat the olive oil in a large skillet, add the pheasant pieces, and cook until golden brown on all sides. Transfer to an ovenproof casserole dish. Sauté the bacon until golden, then add to the casserole. Add the flour, if using, to the pan, stir well and cook for 1 minute until lightly browned. Stir in the apple juice, chicken stock, sage, and parsley, bring to a boil, then pour over the pheasant. Stir in the fennel, season, and bring to a boil on top of the stove. Transfer to a preheated oven and cook at 400°F for 45 minutes, or until just tender. Melt the butter in a small skillet and, when sizzling, add the apple slices and a pinch of sugar, and sauté until golden. Stir into the cooked pheasant, then serve with mashed potato made with crushed garlic, and a glass of warming red wine.

uce that tastes terrific

about ⅔ cup olive oil

2 pheasants,
cut into 8 pieces

6 slices of bacon, cut
into small pieces

1 tablespoon flour
(optional)

⅔ cup apple juice

1¼ cups chicken stock

2–3 large sprigs
of sage

1 large bunch of fresh
flat-leaf parsley

1 fennel bulb,
thickly sliced

2 tablespoons butter

2 apples, sliced

pinch of sugar

salt and freshly
ground black pepper

Serves 4–6

Pasta

Pappardelle
with olive and parsley sauce

A good-quality dried pasta is best for this recipe, though fresh pappardelle from a very good Italian deli could also be used.

To make the sauce, heat 1 tablespoon of the olive oil in a small skillet, add the shallot, and sauté until softened and transparent. Add the remaining sauce ingredients, simmer for 10 minutes, then place in a blender or food processor and puree until smooth. Bring a large pan of salted water to a boil, add the pasta and cook according to the package instructions until just al dente. Drain and serve immediately in heated pasta dishes, with the sauce spooned over, then sprinkled with the chopped fresh oregano.

1 lb. good quality dried pappardelle

4 tablespoons chopped fresh oregano, to garnish

olive and parsley sauce

½ cup olive oil

1 small shallot, finely chopped

⅔ cup green olives, pitted

2 garlic cloves, crushed

4 sun-dried tomatoes, finely chopped

grated zest and juice of 1 lemon

1 bunch of fresh flat-leaf parsley

1 bunch of fresh oregano

salt and freshly ground black pepper

Serves 4

Baked green gnocchi
with creamy onion confit

A warming dish with sweetly flavored onions in a cream sauce, baked with green spinach gnocchi and Parmesan.

To make the onion confit, heat the olive oil in a large pan, add the onions, then stir. Cover the pan with a lid and cook until softened and golden.

Take care not to burn the onions, and add a little more olive oil if necessary.

Add the garlic, cream, and eggs, and heat gently.

Meanwhile, cook the gnocchi in boiling salted water for a few minutes until they rise to the surface.

Drain and stir into the sauce.

Spoon the mixture into a gratin pan, sprinkle with the Parmesan, and cook in a preheated oven at 400°F for 15 to 20 minutes.

Sprinkle with pepper and serve immediately.

1 lb. fresh gnocchi, spinach or plain

2 oz. freshly grated Parmesan cheese

freshly ground black pepper

onion confit

2 tablespoons olive oil

1 lb. onions, finely sliced

1 garlic clove, crushed

⅔ cup light cream

2 eggs, lightly beaten

Serves 4

Pumpkin ravioli
with a pumpkin and dill sauce

Olive oil and chopped dill are used to flavor both the sauce and the filling in this gorgeous, golden ravioli recipe.

1 egg

2 cups all-purpose flour

water, as needed

ravioli filling

2 tablespoons olive oil

1 lb. pumpkin or butternut squash, peeled and finely cubed

1 tablespoon chopped fresh dill

salt and freshly ground black pepper

pumpkin and dill sauce

2 tablespoons extra-virgin olive oil

8 oz. pumpkin or butternut squash, peeled and finely cubed

1 shallot or small onion, finely chopped

2 tablespoons chopped fresh dill

sea salt and freshly ground black pepper

freshly grated Parmesan cheese, to serve

Serves 4

To make the filling, heat the oil in a skillet, and sauté the pumpkin cubes very gently for about 20 minutes until very soft. Add the dill, salt, and pepper.

To make the pasta, mix the egg and flour together, adding enough water to form a stiff but smooth dough. Knead for 10 minutes until very smooth. Cover and rest the dough for 40 minutes. Using a pasta machine, roll out the dough into a long strip about 5 inches wide. Place spoonfuls of filling in 2 rows, about 1 inch from the top and bottom, and the same distance apart, until half the pasta sheet is covered. Moisten the edges and the spaces in between with water. Fold over the other half of the sheet, then flatten and seal the sections between the fillings. Cut out the squares with a zigzag ravioli wheel. Alternatively, use a ravioli cutter or a pasta machine with the ravioli attachment.

Chill the filled ravioli while you make the sauce.

To make the sauce, heat the oil in a skillet, add the pumpkin and shallot, and sauté until tender. Add the dill, salt, and pepper. Cook the ravioli in a large pan of boiling salted water for 4 to 5 minutes, until they rise to the surface. To serve, pour into heated plates, spoon over the sauce, and sprinkle with Parmesan.

a **wonderful**, colorful sauce—served

over ravioli made with this delicious herbed pumpkin filling

Rice

Beet risotto

For fantastic flavor, roast the beetroot yourself. This recipe is also delicious served with a spoonful of sour cream.

8 oz. cooked beets, peeled and diced, or 1 lb. fresh beets, unpeeled

2 tablespoons olive oil

1 small onion, finely chopped

1½ cups risotto rice

1½ cups hot vegetable stock

sea salt and freshly ground black pepper

to serve

chopped fresh sage leaves

shavings of fresh Parmesan cheese

Serves 4

If using fresh beets, roast them, unpeeled, in a preheated oven at 400°F for 30 minutes. Peel and dice. To make the risotto, heat the oil in a saucepan, add the onion and sauté until softened. Stir in the rice, add a ladle of hot stock, stir, and allow the rice to absorb the stock. Stir in the beets and more stock, stir again, and simmer until absorbed. Repeat until you have used all the stock. Taste, adjust the seasoning and serve, topped with chopped sage and shavings of fresh Parmesan.

the very essence of tomato—**intense and concentrated**—

topped with full-flavored basil pesto

Baked tomato risotto
with green pesto

Although there are lots of wonderful flavors for risotto, tomato and pesto are especially good. Whip up your own pesto when basil is plentiful, using the recipe on page 21. Otherwise, use good-quality ready-made pesto. The dense texture of plum tomatoes is best suited to this recipe—and it's simple to bake them in advance, very slowly, in a low oven, so their liquid evaporates leaving an intense, concentrated flavor.

1 lb. ripe plum tomatoes, sliced

1 garlic clove, crushed

2 tablespoons olive oil

1 onion, finely chopped

1½ cups risotto rice

1½ cups hot vegetable stock

salt and freshly ground black pepper

to serve

4 tablespoons basil pesto (see page 21)

shavings of fresh Parmesan cheese

Serves 4

Place the tomatoes in a roasting pan with the garlic, sprinkle with salt and freshly ground black pepper, then bake in a preheated oven at 325°F for 1 hour. Heat the olive oil in a saucepan, add the onion and lightly sauté until golden. Stir in the rice, add a ladle of hot stock, then stir and allow the rice to absorb the stock. Add more stock, stir, and simmer until absorbed. Repeat until you have used all the stock. Taste and adjust the seasoning. Serve, topped with the baked tomatoes, a spoonful of basil pesto and some shavings of fresh Parmesan.

Vegetables

Mediterranean ragoût
of marinated vegetables

Choose your own selection of vegetables, according to what's in season. Baby artichokes and fava beans are especially delicious additions.

Blanch the zucchini in a pan of boiling, salted water. Remove with a slotted spoon and set aside. Add the onions and boil until soft and tender, then remove with a slotted spoon and set aside.

Blanch the mushrooms in boiling water, then drain. Place the marinade ingredients in a large pan, bring to a boil, add the blanched vegetables and simmer for 10 minutes. Cool, transfer to a bowl, cover, and refrigerate for a few days for the flavors to develop. Drizzle with extra-virgin olive oil and serve, accompanied by crusty bread.

- 1 lb. small zucchini
- 8 oz. pearl onions
- 2 cups button mushrooms
- extra-virgin olive oil, to serve

marinade

- 2½ cups water
- juice of 3 lemons
- 1¼ cups olive oil
- 2 sprigs of thyme
- 1 bunch of fresh flat-leaf parsley
- 4 bay leaves
- 1 small celery stalk, with leaves
- 10 black peppercorns
- 10 green olives

Serves 4

Pickled eggplants
for bruschetta or pasta sauce

A useful bruschetta topping or pasta sauce, sprinkled with chopped sun-dried tomatoes.

- 1 shallot
- 2 fresh bay leaves
- 2 teaspoons salt
- 1¼ cups vinegar
- 2½ cups water
- 2 lb. baby eggplants (or large ones cut into chunks)
- 8 whole garlic cloves
- 2 fresh red chiles
- 4 sprigs of rosemary
- olive oil, to cover

Place the shallot, bay leaves, salt, vinegar, and water in a pan and bring to a boil. Add the eggplants. Put a plate on top to keep them submerged, simmer for 10 minutes until soft, then drain. Pour boiling water over two 1-quart Mason jars and place in the oven at 300°F for 5 minutes. Remove, drain, and pack in the eggplants, garlic, chiles, and rosemary. Cover with oil, leaving ½ inch headroom. Seal, set in a roasting pan filled with water, and place in the oven at 300°F for 15 minutes, or in a boiling water-bath canner according to USDA guidelines. Cool, chill for 2 days before eating, and use within 1 month.*

Makes two 1-quart jars

Roasted bell peppers
marinated in lemon-flavored oil

A great bell pepper recipe—ideal on its own, on bruschetta or with pasta.

- 8 red or yellow bell peppers
- peeled rind of 1 lemon
- about 1¼ cups virgin olive oil

Broil the bell peppers on all sides until the skins are blistered and blackened. Wrap in plastic wrap for 3 to 5 minutes. Slip off the skins, pull out the stems and seeds, cut in half, and place in 4 glass jam jars. Add the lemon peel, then pour over olive oil to cover. Chill 2 to 5 days before using and use within 1 week.

Makes four ½-pint jars

*Please refer to the directive on page 4.

Spiced garbanzos
with eggplant and tomatoes

Olive oil, used as a braising medium, gives the garbanzos loads of flavor. If you're short of time, canned garbanzos make this Middle-Eastern-influenced dish very quick and easy to prepare. Serve with saffron rice and dishes such as pan-fried lamb fillet.

6 tablespoons extra-virgin olive oil

1 onion, finely chopped

1 garlic clove, crushed

½ teaspoon turmeric

small pinch of ground coriander

small pinch of ground cumin

1 eggplant, unpeeled, diced

1 lb. plum tomatoes, blanched, peeled, seeded and finely chopped

2 tablespoons red wine or stock

1½ cups garbanzo beans, freshly cooked or canned

sea salt and freshly ground black pepper

Serves 4

Heat the olive oil in a large skillet, add the chopped onion and the garlic, and sauté gently until softened and transparent.

Stir in the remaining ingredients, bring to a boil and simmer gently until the tomato sauce has slightly thickened. Serve immediately.

a great main dish for vegetarians—and also **delicious** as an appetizer

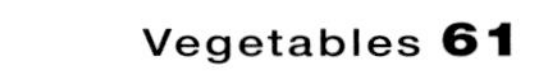

Orange and lemon
polenta cake

Olive oil can be used to replace butter in cakes, and the type of oil you use will create the flavor. If you don't want it too strong, start with a light oil and move on up to a strong Spanish or Greek oil. Polenta gives the cake a lovely, nutty texture and a pretty green-gold color. This cake is wonderful served plain with coffee, or as a dessert with crème fraîche and a glass of sweet white wine.

. . . and so

Place the orange and lemon in a saucepan with cold water to cover. Bring to a boil and simmer for about 30 minutes until very soft. Drain and reserve the fruit, then cool. Place in a blender or food processor and puree until smooth.

Beat the eggs with an electric mixer until pale and fluffy (about 5 to 10 minutes), then whisk in the sugar. Mix the flour with the baking soda and salt, stir into the egg mixture, then mix in the olive oil. Fold in the polenta and the pureed fruit.

Pour the batter into an 8-inch springform pan. Cook in a preheated oven at 350°F for about 50 minutes, until a skewer, inserted into the center of the cake, comes out clean. Remove from the oven, place on a wire rack, sprinkle with confectioners sugar and, when cool, remove from the pan. Serve plain, or with whipped cream or crème fraîche.

ng sweet

1 orange

1 lemon

4 eggs

½ cup sugar

1¾ cups
all-purpose flour

3 teaspoons
baking soda

pinch of salt

1¼ cups pint olive oil

⅔ cup polenta or
yellow cornmeal

to serve

confectioners sugar

4 tablespoons
crème fraîche,
(optional)

Serves 8–10

Index